Guenson EXIL

Economic profitability of vegetable farms

Guenson EXIL

Economic profitability of vegetable farms

Analysis of the economic profitability of vegetable farms in the commune of Saint Raphaël (2022)

Imprint
Any brand names and product names mentioned in this book are subject to trademark, brand or patent protection and are trademarks or registered trademarks of their respective holders. The use of brand names, product names, common names, trade names, product descriptions etc. even without a particular marking in this work is in no way to be construed to mean that such names may be regarded as unrestricted in respect of trademark and brand protection legislation and could thus be used by anyone.

Cover image: www.ingimage.com

This book is a translation from the original published under ISBN 978-620-6-72569-5.

Publisher:
Sciencia Scripts
is a trademark of
Dodo Books Indian Ocean Ltd. and OmniScriptum S.R.L publishing group

120 High Road, East Finchley, London, N2 9ED, United Kingdom
Str. Armeneasca 28/1, office 1, Chisinau MD-2012, Republic of Moldova, Europe
Printed at: see last page
ISBN: 978-620-8-22231-4

ANALYSIS OF THE ECONOMIC PROFITABILITY OF VEGETABLE FARMS IN THE COMMUNAL SECTION OF SAN-YAGO, COMMUNE OF SAINT-RAPHAËL (2022)

DEDICATION

This work is dedicated to :

- ➢ *My parents Bertilde Faustin and Etienne EXIL;*
- ➢ *My brothers Bermano and Levity ;*
- ➢ *Administrator Mielleda LORISTHENE ;*
- ➢ *My colleagues in the class of 2017-2022.*

ACKNOWLEDGEMENTS

This dissertation is the fruit of considerable effort, and it is essential to express my gratitude to those who have supported me in various ways to bring this work to fruition.

*First of all, special thanks go to my scientific advisor, Agroeconomist **Désilhomme SATYR M.Sc**, who, despite his many commitments, generously offered his advice and comments from the beginning to the end of this work.*

*I would also like to express my gratitude to **Jackson GERVAIS** Ing-Agr. M. Sc for his invaluable advice, and to the economist **Maqueinder CHARLES** PhD student, whose active contribution enriched my training.*

*Many thanks to the university's **rectorate** for its commitment and ongoing support for student education and development.*

*I would also like to express my gratitude to the **Scientific Committee** for its commitment to improving the quality of research at the university.*

Many thanks to those in charge of the deanship, in particular the FSAA coordinator.

***Jackenson MAURICETTE** Ing. Agr. M. Sc, for their ongoing support.*

*A fist of thanks also goes to all my comrades in the class of 2017- 2022 especially to: **JOSEPH Yvenel, BELFLEUR Roodley, RAPHAEL Valmir, CHERY Dieuseul** for their support.*

*We would like to extend our warmest thanks to agricultural engineers **Lordanson JOACHIM** and **Willy PARFAIT**, as well as student **Charlot JOSEPH**, for their invaluable assistance.*

*I would also like to thank **all those who opened their doors to me** so that I could carry out the survey.*

*Finally, I would like to thank **the members of the jury**, who did their best to be present, and who have our deepest gratitude.*

SUMMARY

The commune of Saint-Raphaël plays a crucial role in vegetable production in the northern region, but it is facing problems that are jeopardising its economic profitability. In addition, the municipality lacks comprehensive data on the profitability of vegetable farms. With this in mind, our research set out to analyse the economic profitability of vegetable farms in the aforementioned commune, with the aim of identifying the main factors influencing their profitability. To this end, five specific objectives were set. Data were collected from a randomly selected sample of 113 vegetable farms and analysed using specialised software such as SPSS 17.0 and InfoStat. The results show that 80% of those surveyed were men, with an average age of 45, and that 71% of the vegetables produced were destined for the market. With regard to the categorisation of vegetable farms, 11 farms, or 10%, were considered poor, 18 farms, or 16%, were classified as poor, 52 farms, or 46%, were categorised as average, while 32 farms, or 28%, were considered well-off. In addition, vegetable farms generate a total income of 674,767.25 HTG ($5,111), with an estimated economic rate of return of 50%. However, leek growing has a higher rate, at 61%. At the end of the study, it was recommended that market gardeners focus more on growing leeks, strengthen hydro-agricultural infrastructure, improve their entrepreneurial skills and develop agri-food businesses to ensure the sustainability of vegetable products.

Keywords: vegetable farm, economic profitability, leek, beetroot, carrot and Saint Raphaël

TABLE OF CONTENTS

I- INTRODUCTION

Agriculture is the main source of income for 80% of the world's poor, playing a crucial role in reducing poverty and improving food security (World Bank, 2021). In 2018, it accounted for 4% of global GDP, and even exceeded 25% in some developing countries. This is particularly the case in Haiti, where agriculture accounts for 22% of GDP and creates 68% of total national employment (Jeanniton & Bellande, 2016). In 2021, average world vegetable production is estimated at 230,882 metric tonnes, broken down into 903.19 tonnes in Asia, 89.27 tonnes in Europe, 85.6 tonnes in America, 73.14 tonnes in Africa and 3.4 tonnes in Oceania (Statista, 2023). In Haiti, average vegetable production was estimated at 155,200 tonnes in 2013, falling to 120,748 tonnes in 2020. (FAOSTAT, 2022).In fact, market gardening plays a major role in several regions of the country, particularly in communes such as Kenscoff, Maïssade, Cerca-la-Source, Saint-Raphaël, Ouanaminthe, Capotille, Hinche and others (St-Pierre, 2022). These areas are characterised by a wide variety of crops, including aubergine, calalou, shallots, leeks, beetroot, tomatoes, mirliton, local cabbage and many others.In Saint-Raphaël, market gardening plays a particularly important role in the economic life of farmers. The main vegetables grown include onions, carrots, leeks, tomatoes, aubergines and beetroot (François, 2017). These crops are grown to meet the food, educational and economic needs of the local farmers.Consequently, given the crucial importance of vegetable crops in this commune, improving their economic profitability must be a central objective of development policies. The aim of this study is to analyse the economic profitability of the main vegetable farms in Saint-Raphaël in 2022, in order to identify the factors influencing their profitability and determine their level of profitability. To this end, this study is organised into four chapters: the first introduces the general context and the problem, the second discusses the literature reviews, the third sets out the methodological framework of the study and the fourth presents the results and discussions, particularly on the economic profitability of vegetable farms in the commune of Saint-Raphaël.

1.1- Issues

Global agricultural production is threatened by the increasing effects of climate change, particularly in regions of the world that are already suffering from food insecurity (World Bank, 2021). Despite the importance of agriculture to the economy, the slow increase in food production and marked fluctuations from one year to the next remain major and chronic problems for developing countries and are the main causes of their worsening poverty and food insecurity

(FAO, 2000).In Haiti, the majority of farmers practise subsistence farming due to a lack of technical support, and are largely dependent on traditional cultivation techniques (FAO, 2022). 90% of farms depend on rainfall. However, only 10% are irrigated, but they are faced with problems of sedimentation and poor drainage (IFAD, 2021). This is typical of the commune of Saint-Raphaël, where the irrigated areas are in poor condition due to drought, lack of canal cleaning and poor management of the hydro-agricultural infrastructure (François, 2017). These constraints are having a negative impact on vegetable farms in the area, thereby hampering the technical and economic profitability of vegetable crops.

In addition to the problems of the unavailability of hired labour and irrigation water, the lack of assistance (technical, material and financial), the proliferation of bio-aggressors (disease, insects and weeds), and the high cost and unavailability of mineral fertiliser. Other structural problems are also apparent, such as the lack of agricultural credit combined with farmers' low self-financing capacity, the limited accessibility and insufficient availability of improved agricultural inputs, combined with inadequate agricultural support services, are also problems facing farmers in the commune of Saint-Raphaël.

On the other hand, despite the very high yield potential of vegetable crops, the aforementioned problems compromise their yield in the commune, including leek (1,600 kg/ha), carrot (3,000 kg/ha), beetroot (1,546 kg/ha) and cabbage (5,000 kg/ha), (MARNDR, 2015). These yields translate into an insignificant average added value, reaching only 128,490 HTG/ha in 2015 (Ibid.). Moreover, the economic profitability of production vegetable growing, in particular the crops covered in the study, has not been sufficiently documented in previous studies carried out in the study area. In view of these problems, it is appropriate to ask: What are the main factors influencing the economic profitability of vegetable crops in the municipality of Saint-Raphaël? What are the main constraints on economic profitability in the commune of Saint-Raphaël? Aware of these problems and keen to contribute to solving them, this work aims, on the one hand, to identify the factors that influence the economic profitability of vegetable crops and the constraints associated with vegetable production. Secondly, it seeks to propose solutions for improving the economic profitability of vegetable farms in the commune of Saint Raphaël. These are the main concerns of this study, which aims to analyse the economic profitability of vegetable farms in the commune of Saint-Raphaël.

1.2- Research hypotheses

In the context of this work, two hypotheses are being tested.
➢ Socio-economic factors influence the economic profitability of vegetable growing in the commune of Saint-Raphaël.
➢ Technical, environmental, infrastructural, economic, financial and institutional constraints have a negative impact on economic profitability in Saint-Raphaël.

1.3- Aims of the study

This study includes a general objective as well as specific objectives.

1.3.1- General objective

The general aim of this study is to analyse the economic profitability of vegetable farms in the commune of Saint-Raphaël.

1.3.2-Specific objectives

The specific objectives of this study are to :

➢ Analyse the socio-economic characteristics of vegetable growers and farms in the municipality of Saint-Raphaël ;
➢ Identify the technical itineraries used by market gardeners in the commune of Saint-Raphaël ;
➢ Categorising vegetable farms in Saint-Raphaël ;

➢ Evaluate the economic profitability indicators for each crop in the municipality of Saint-Raphaël ;
➢ Identify and analyse the constraints and strengths of vegetable farms in Saint-Raphaël.

1.4 Interest of the study

This study is of threefold personal, scientific and technical interest:

➢ For its own sake, this is an academic requirement imposed by the faculty with a view to obtaining our final degree;
➢ Its scientific value lies in the fact that it highlights qualitative, quantitative and verifiable data on the economic profitability of farms in the target area, which can provide current scientific literature with additional information that could serve as a compass for further research in the study area and elsewhere;
➢ Because of its technical interest, the results of our study could, on a local

scale, enable market gardeners to make sound economic and financial decisions and contribute as a guidance and consultation tool to the agricultural development policy of the municipality of Saint-Raphaël.

1.5- Limitations of the study

The aim of the study was to analyse the economic profitability of vegetable farms in the commune of Saint-Raphaël. However, it did not take into account yield, the agricultural profit generated by each type of crop, environmental performance indicators, the impact of pests and diseases, and yield losses due to bioaggressors and the different vegetable growing systems in the municipality.

II- REVIEW OF LITERATURE

This chapter reviews the key concepts, such as the notion of farm, cropping system, profitability and its determinants in the theoretical literature. The soil and climate requirements of the crops considered are then presented.

2.1- Definition of key concepts

This section provides definitions of key concepts related to the study, drawn from a range of authors.

2.1.1- Farming operations

Several definitions of the farm have already been proposed by different authors. Some emphasise the systemic dimension. For example, the FAO (2001) believes that the farm should be studied as a system made up of several interacting elements (FAO, 2001).

For Chombart de Lauve et al (1963) quoted by Jimbira (2004): The farm is an economic unit in which the farmer practices a production system with a view to increasing his profit (Chombart de Lauve, Poitevin, & Tirel, 1963). This approach assimilates the farm to a business and the farmer to an entrepreneur.

For Laurent C. and Rémy J (2000):
"A farm is a social construct with multiple dimensions: spatial, agronomic, economic, statistical, institutional, symbolic...". (Laurent & Rémy, 2000).For the FAO (1995), a farm is an economic unit of agricultural production, managed by the head of the farm, including the animals and the land farmed, regardless of the type of tenure and the quantity sown (FAO, 1995).

2.1.2- Growing system

Several authors have contributed to the definition of the cropping system concept. Sebillotte (1993), quoted by Malézieux & Trébuil (2000), defines a cropping system as "a set of technical methods used on plots of land that are treated in an identical manner" (Sebillotte, 1993). According to his understanding, it is defined by three important elements. These are :

➢ Types of crops grown ;
➢ Order of appearance of cultures ;

➢ The technical itineraries used, including the choice of cultivars.
For Gras (1990) quoted by (Jouve, 2003). A cropping system is any plot of land

that is sown and then treated in a homogeneous way, in terms of the crops grown, their order of succession and the technical itineraries used (Gras, 1990).

2.1.3- Technical itinerary

For Sebillotte (1993) quoted by (François, 2008), the technical itinerary is defined as a methodical and organised sequence of techniques aimed at sowing an area in order to optimise production (Sebillotte, 1993).

2.1.4- Vegetable farming

A vegetable farm is any farm that has produced vegetables, regardless of the area sown to vegetables or the technical orientation of the farms (Agreste, 2013). In other words, it is a farm that has declared that it grows fresh vegetables.

2.1.5- Vegetables

Vegetables are defined as any edible part of a vegetable species. The main cultivated vegetables can be classified according to the organ consumed (Yehouenou, 2011). For Bognini (2010), vegetables are herbaceous plants whose edible parts are harvested from the plant while it is still standing or during its resting period (Bognini, 2010). Others define vegetables as the fresh parts of plants that are eaten on their own, as food supplements or as side dishes. Thus, the main vegetables grown can be classified according to their nature, their demand on the market and where they are grown (Yehouenou, 2011). The following table shows the different types of vegetables classified according to the part consumed.

Table 1: Classification of vegetables according to the type of organ consumed

Fruit vegetables	tomatoes, aubergines, peppers, okra, melons, cucumbers and others.
Leafy vegetables	lettuce, amaranth, lamb's lettuce, celery, cabbage, spinach, fennel, sorrel and others.
Root vegetables	carrot, beetroot, turnip, radish and others.
Stem vegetables	asparagus, leeks, alliaceae bulbs: garlic, shallots, onions and others.
Flowering vegetables	cauliflower, broccoli, capers, artichokes and others.
Tubers	yams, potatoes and others.
Herbs	chervil, chives, tarragon, bay leaves, parsley, thyme and others.

Source: (Jean-Denis, 2015) **quoted by** (St-Pierre, 2022)

2.1.6- Economic profitability

According to Pirou, economic profitability is none other than the capacity of capital placed in any institution, or invested, to generate income (Pirou, 2005).

Economic profitability measures the efficiency of a single production process, independently of the financial, financing and tax elements t h a t are exogenous to it. It is expressed as a percentage, and the aim of this approach is to compare the variation in its numerator with that in its denominator, i.e. when the numerator grows faster than the denominator, economic profitability increases (Durant, 2005).

For Lassana et al (2021), profitability remains a sine qua non but not sufficient condition to guarantee the survival of a business (Lassana et al, 2021). Economic profitability measures the ratio between the income generated by a business and the total capital (equity + financial debt) released to obtain it (Barbacar et al, 2020).

2.1.7- Gross revenue

The Gross Product is all that is produced by the farm, i.e. all t h e values it has generated as part of its current professional activity (Guen, 2016). It is the annual value of production, whether this production is sold or partly consumed by the farmer and his family. The PB depends on the yields obtained and the value of production (CIRAD, 2009).

2.1.8- Intermediate consumption

Intermediate consumption corresponds to goods and services consumed entirely during the year: seeds (purchased or produced by the farmer), various inputs (fertilisers, pesticides) and services provided by agents outside the farm (El Ouaamar et al, 2019).

2.1.9- Added value

Value added measures the creation of wealth intrinsic to the production process (excluding any subsidies received). It is equal to the difference between the value produced (the gross product, including the self-consumed part of this product) and the value of the goods and services consumed in whole or in part during the production process (CIRAD, 2009).

2.1.10- Valuation of the working day

Valuing the working day is a way of recognising and appreciating the time spent by a farm worker. It is calculated by dividing total production expressed in monetary terms by the number of days worked. This indicator is of crucial importance and is useful for analysing the performance of different crops (CIRAD, 2009). In addition, the net value of the working day corresponds to the net value added divided by the number of working days.

2.1.11- Farm income

Agricultural income is defined as the income generated by agricultural activities during a given accounting period, even if the corresponding revenues are in some cases not received until later (El Ouaamar et al, 2019).

2.1.12- Economic rate of return

The economic rate of return measures the ratio of income to costs expressed in value. It measures agricultural income per unit of capital invested. Thus, the economic rate of return expresses the total gain obtained by investing one monetary unit (Lassana et al, 2021).

2.2- Nutritional importance of vegetable crops

Vegetables include leaves, fruit, roots and flowers. The term vegetable is useful in nutrition and everyday terminology. They play an important role in the diet. They are almost all rich in carotene, vitamin C and vitamin A, rich in antioxidants and contain significant quantities of calcium, iron and other minerals. Their B vitamin content is often low. They are low in energy and protein. They contain a large proportion of substances that cannot be broken down by the digestive enzymes, which will be added to the faeces. However, green vegetables are a good source of vitamin A and C, and tropical leafy vegetables are a good source of minerals (Ca, Iron) and vitamin A (FAO, 2004).

2.3- Vegetable production in Haiti

Fresh vegetable production in Haiti is estimated at 121,400 metric tonnes (Moïse, 2017). The most recent FAO data on vegetable production in Haiti dates back to 2020, indicating that around 26,000 hectares were devoted to vegetable cultivation, with an estimated production of 120,748 tonnes. Production has fallen since 2013, from around 125,000 tonnes to around 120,000 tonnes in 2020, affecting various crops such as cabbage, lettuce, tomatoes, leeks, chillies, peppers, spinach, potatoes, carrots and other species (FAOSTAT, 2022).

2.4- Vegetable production zone in Haiti

Vegetables are grown in almost all high-altitude areas and in a few low-altitude areas in Haiti. The types of crops grown differ from one region to another, depending on the specific soil and climate conditions in each area and the demand for vegetables. The table below shows the different crops grown in the various regions of the country.

Table 2: The main vegetable production areas in Haiti and the crops grown

Department	Zone	Culture
West	Plaine de l'Arcahaie, Plateau des palms, Cornillon plateau	Aubergine, cabbage leek beetroot tomato and others.
West-South-East	Kenscoff-Seguin and Forêt des Pines	Leeks, onions, cabbage, carrots, potatoes earth, lettuce and others.
South-East	Plateau La Vallée de Jacmel	Cabbage, mirliton, cives, thyme, pepper and others.
Artibonite	Plateau de Goyavier, Plaine des Gonaïves, artibonite valley	Tomato, onion, lalo, okra, aubergine, cabbage and others.
Nippes	Salagnac plateau	Carrot, cabbage, local leek, parsley, mirliton and others.
North	Saint-Raphaël	Onions carrot amaranth beetroot, leek and others
North-east	Plateau Carice-Mont Organisé Plaine Maribaroux	Cabbage, tomato, pepper and others.
South	Cayes-Cavaillon plain	Tomato, amaranth, leek
Centre	Bas plateau, Plateau de Baptiste	Tomato, cabbage, chilli and others
Northwest	La Croix St-joseph	Cabbage and carrot

Source: Bellande quoted by (St-Pierre, 2022)

2.5- Presentation of the crops examined in the study

In this study, three vegetable crops were selected: leeks, carrots and beetroot. In fact, each crop has its own pedoclimatic requirements in order to express their potential. The following paragraphs set out the soil and climate requirements of the crops covered in the study.

2.5.1- Carrot cultivation

The carrot, a member of the Apiaceae family with the scientific name Daucus carrota, has many varieties distinguished by the shape of their root and their production period. There are long, semi-long and semi-short varieties.

2.5.1.1-Temperature

Carrots thrive in cool climates. The optimum temperature for seed germination is 18°C, with a minimum of 7°C. For growth, a temperature between 20 and 27°C is ideal, while the best root colour is obtained when the temperature is between 16 and 21°C. This temperature range should be maintained for around three weeks before harvesting (Adamou, 2020).

2.5.1.2-Soil requirements

Growing carrots does not require high-quality soil, but it is best to avoid stony soils to ensure straight roots. The ideal soil is sandy loam. Carrots cannot tolerate salty or acidic soil, or irrigation water. The optimum soil pH is between 6.5 and 7 (Adamou, 2020).

2.5.1.3-Water requirements

A regular, uninterrupted supply of water is essential for carrots, especially during the growing period, to prevent cracking and deformation of the roots. On average, carrots require around 350 mm of water (Adamou, 2020).

2.5.2- Leek cultivation

Leeks, which belong to the Liliaceae family and have the scientific name Allium porum, stand out for their high resistance to pests and diseases. There are a multitude of varieties, each with its own characteristics and advantages.

2.5.2.1-Temperature

Leeks thrive in a mild, damp climate, but some varieties are highly resistant to cold. It grows best between 15 and 25°C, with a vegetative phase down to 2°C (ITCMI, 2022).

2.5.2.2-Soil requirements

Leeks prefer deep, well-aerated soils rich in organic matter, with a pH of between 6.5 and 7. It does not like soils that are too chalky (pH>8) (ITCMI, 2022).

2.5.2.3-Water requirements

The water requirements of leek crops are considerable and vary according to the degree of soil cover. On average, these requirements are estimated at between 300 and 400 m^3 / ha (ITCMI, 2022).

2.5.3- Beet growing

Beetroot, a member of the Chenopodiaceae family with the scientific name Beta vulgaris, is grown all over the world in different varieties, characterised by their shape and size.

2.5.3.1- Temperature

Beetroot is sensitive to frost, with temperatures of between -3 and -7°C, depending on the duration and stage of growth. The optimum soil temperature for sowing is between 5 and 8 °C, and that the optimum growing temperature is between 18-25 °C, but cold spells (<5 °C) for two to three weeks after sowing can stunt its growth (Agridea, 2017).

2.5.3.2-Water requirements

Beetroot prefers moderate rainfall and tolerates drought well in deep, well-structured soils. Their average water requirement is between 600 and 700 mm (Agridea, 2017).

2.5.3.3-Soil   requirements

An adequate supply of calcium promotes soil stability and reduces the risk of diseases such as black foot. Beetroot thrives in sandy soil, rich in organic matter but poor in humus, with a pH of around 6.5 (Agridea, 2017).

2.6- Empirical evidence on the determinants of economic profitability

Early work on the sources of profitability in the agricultural sector focused on farmer training. According to Stefanou and Swati (1988), various types of training can help agricultural producers to increase their profitability. At the level of individual farms, the determinants of profitability may be: gender, size, contact with agricultural extension, managerial capacity and access to credit (Malla & Yabi, 2023). Rodgers (1994), on the other hand, insists on know-how; for him, the market or technological innovation is not the main cause of low farm profitability, but rather the need to improve human capital. Hanson (2008) suggests that farm profitability can be increased by focusing on know-how and increasing farm size.Rajendran et al (2015) measuring the technical performance of farm households producing traditional vegetables conclude that strengthening farmers' associations to encourage knowledge sharing and strengthening relationships between farmers can help improve technical performance (yield). Singbo et al (2014), who estimated the production and sales function of vegetable producers in Benin, concluded that technical performance is influenced by the production environment and extension services. It should be noted that credit can have a positive influence on farm performance if the funds obtained by farmers are used to purchase inputs. Richter et al (1994) state that the lack of cash and credit limits farmers' resources (Albouchi & Bachta, 2007). According to Afful (1987) increasing agricultural productivity requires investment in the land itself.

III-METHODOLOGICAL FRAMEWORK

This chapter begins with a detailed presentation of the municipality of Saint-Raphaël. Secondly, the various methodological phases and the procedures for calculating the economic and financial indicators are presented.

3.1- Physical setting of the study area

The municipality of Saint-Raphaël covers a total area of 183.8 km^2 and is part of the arrondissement of the same name in the department of Nord. It comprises around thirteen (13) localities and fifty-four (54) homes. It is located 48 kilometres from Cap-Haïtien and at an altitude of 366 m (François, 2017). It is bordered by :

➢ In the North via the Grande Rivière du Nord and DonDon ;

➢ To the south by Maïssade ;

➢ To the east via Saint-Michel de l'Attalaye ;

➢ To the west by Bahon and de Pignon respectively.

3.1.1- Administrative subdivision of the municipality of Saint-Raphaël

It is subdivided into four communal sections:

➢ 1ère section communale : Bois Neuf ;

➢ 2ème section communale : Mathurin ;

➢ 3ème section communale : Bouyara ;

➢ 4ème communal section: San-Yago.

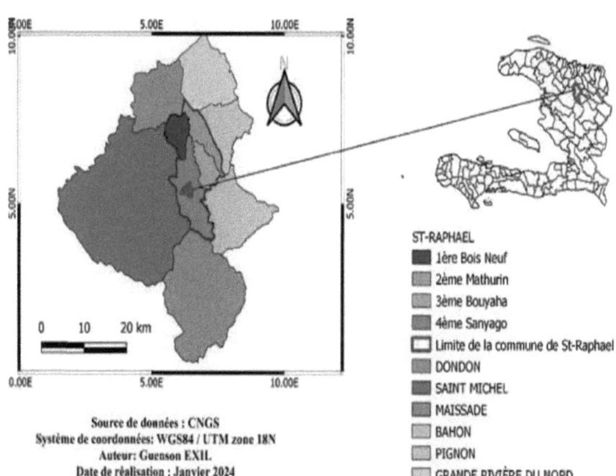

Figure 1: Administrative division and administrative subdivision of the study municipality

3.1.2- Demographics

The total population of Saint Raphaël is estimated at around 53,755, comprising 26,608 women and 27,147 men. 32.6% of the commune's population live in urban areas (4^e section San-Yago), with a population density of 292 inhabitants/km^2 . The communal section of San-Yago covers an area of 98.3 km^2 and has an estimated total population of 37,515, of which 19,965 live in rural areas (10,069 men compared with 9,896 women) and 17,550 live in urban areas (8,723 men compared with 8,827 women). It should be noted that the rural area has 4205 households compared to 3930 in the urban area (IHSI, 2015).

3.1.3- Education

According to MENFP (2011), the commune of Saint-Raphaël includes seventy-five (75) educational institutions. Of these, there are twenty (20) pre-schools, fifty (50) primary schools and five (5) secondary schools (MEF, 2021).

3.1.4- Presentation of the commune's economic activities

This section presents the different activities in Saint-Raphaël by sector.

3.1.4.1- Primary sector

In Saint-Raphaël, agriculture is the main economic activity in which residents of the section engage. Agriculture is practised intensively due to the availability o f irrigation water in most of the area. However, there are two irrigated perimeters in the section, the Merlaine perimeter known as the small perimeter with around 400 farmers and a large perimeter with 6,000 farmers for a sown area of 1,000 ha (MARNDR, 2015).

3.1.4.1.1- Market gardening system in the commune

In rotation with rice, vegetables are grown in the lightest soils with good drainage (MARNDR, 2015). The main vegetable species found in the commune of Saint-Raphaël are: leek, onion, chilli, beetroot, carrot, tomato and cabbage. The table below shows the distribution of vegetable crops in irrigated areas. On the other hand, the most common vegetable crop associations found on the perimeters are :

➢ Pepper, onion, beetroot, calalou ;
➢ Tomato - carrot - cabbage ;

➢ Beetroot, leek, chilli and lime.

Table 3: Cropping calendar for vegetable production in the study area

Species	Jan.	Feb.	March	April	May	June	July	August	Seven.	Oct.	Nov.	Dec.
Carrot									Visit	Visit	Visit	Re
Cabbage									Pe	Visit	Visit	Re
Leek	Visit	Re	Re							Pe	Pe	Visit
Onion	Visit	Visit	Visit	Re	Re					Pe	Pe	Visit
Beetroot	Visit	Re									Pe	Visit
Chilli	Visit	Visit	Visit	Re	Re	Re					Pe	Visit
Tomato	Visit	Re	Re	Re							Pe	Visit

Source:(MARNDR, 2015)

Legend: Pe= Nursery **Se**= Sowing/transplanting En=MaintenanceRe=Harvest

3.1.4.2- Secondary sector

Around 92% of farms are involved in non-agricultural activities in Saint-Raphaël. The secondary sector is not well represented in the communal section of San-Yago, there is a guild that processes sugar cane into syrup and clairin, bakeries that process wheat flour into bread and others (MARNDR, 2015).

3.1.4.3- Tertiary sector

The tertiary sector is represented by the marketing of agricultural products such as vegetables, maize and beans, and agricultural inputs. Charcoal is also sold. In terms of communications, Digicel and Natcom operate in the Section. The urban market operates every day, and the regional market on Thursdays. There is a market in all the communal sections, including San Yago, which operates every Thursday and is located on the road leading to Saint Michel de l'Attalaye (MEF, 2015).

3.1.5- Biophysical aspects of the municipality of Saint-Raphaël

This section describes the local soil and climate, including relief, soil type, temperature and other factors.

3.1.5.1- Relief

The commune of Saint Raphaël is made up of plains and mountains. The Bois neuf and Mathurin sections are very hilly, with an average gradient of 90%. The Bouyaha and San-Yago sections, on the other hand, are less rugged (François, 2017).

3.1.5.2- Type of soil

The irrigated area of Saint-Raphaël is characterised by brown calcareous soils and vertisols, the latter having a high swelling clay content that favours the adaptation of rice cultivation. However, along the river and the perimeter, the soils are very deep with a silty-sandy texture and a good organic matter content (MARNDR, 2015).

3.1.5.3- Temperature

The average maximum monthly temperature in the municipality is 30^0 C, as the hottest months of the year are July and August, with an average temperature of 32^0 C. On the other hand, the average minimum monthly temperature is 20.2^0 C (see figure below). The average annual temperature in the area is 24.16^0 C (NAZA, 2016).

Figure 2: Monthly temperature distribution for the municipality of Saint-Raphaël

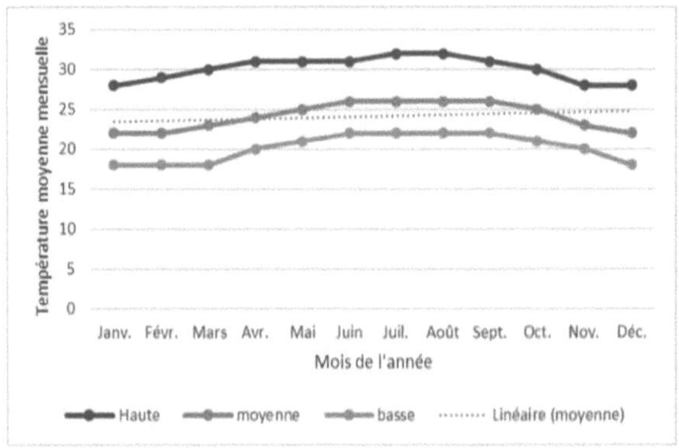

Source: author's construction based on MERRA-2 satellite-era data

3.1.5.4- Rainfall

In Saint-Raphaël, the average monthly rainfall is 65.49 mm, with February considered the driest month. On the other hand, September saw the highest rainfall of the year, with an average of 120.2 mm. All in all, the municipality receives 785.9 mm of rain, as the rain that actually contributes to feeding aquatic environments and recharging groundwater is 687.9 mm (NAZA, 2016). The following figure shows the distribution of rainfall and effective rainfall in the study area.

Figure 3: Distribution of average monthly rainfall in the study area

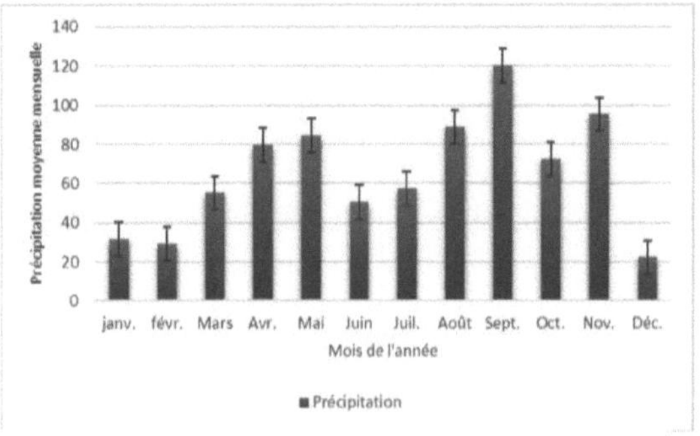

Source: author's construction based on MERRA-2 satellite-era data

3.1.5.5- Water resources

The Saint-Raphaël area has a relatively dense hydrographic network. This network is dominated by the Bouyaha river, which originates at Marmelade, then meanders through the commune of Dondon before arriving at Saint-Raphaël. There are two irrigated areas in the San-Yago section. These are: the large area starting from the dam on the Bouyaha river and the small Merlaine area or Merlaine system (François, 2017).

3.2- Data collection equipment and tools

Various materials were used to carry out this work. These include

3.2.1- Teaching materials

➢ Books and course documents ;
➢ Electronic documents ;

➢ Periodical articles ;

➢ Telephone and computers.

3.2.2- Collection and writing materials

Various types of equipment were used to collect and type data in the field:

➢ Pen, pencil, notebook and paper: for data collection ;

➢ Computer: for research and writing the dissertation;

➢ Survey form: for data collection ;

➢ Digital camera: for taking photos.

3.2.3- Analysis tools

The following software was used to analyse the data:

➢ OpenEPI and Statrek Version 3: To set the sample size and select individuals at random without replacement;
➢ Excel version 2019: to build the database, create figures and some tables;
➢ QGIS version 3.34: for producing the map and defining the boundaries of the study area;
➢ SPSS 17.0.0: for monovariate analyses ;

➢ InfoStat version 2018: for analysing economic profitability indicators for vegetable crops.

3.3- METHOD

Research methodology refers to the procedures or techniques used by the researcher to carry out the study. To this end, this work was carried out according to well-defined methods, comprising the following main phases:

➢ Phase I: Desk study ;

➢ Phase II: Exploratory survey ;

➢ Phase III: sampling ;

➢ Phase IV: Questionnaire design ;

➢ Phase V: Data collection ;

➢ Phase VI: processing and analysing the data and writing up the report.

3.3.1- Documentary study

In this phase, scientific journals, dissertations, books, reports, course notes and other relevant documents on the research topic and the study area were consulted. These documents were used to develop the theoretical and conceptual framework (literature review) and to present the study area and discuss the results obtained.

3.3.2- Exploratory survey

This phase enabled the secondary information gathered to be completed and facilitated the development of the sampling plan for data collection in the field. It also provided an opportunity to observe the economic, socio-cultural and agronomic conditions in the area, and to talk to local farmers to identify the various problems encountered in vegetable production.

3.3.3- Sampling

In this study, a simple random probability sampling method was used. The sampling frame consisted mainly of market garden holdings in the communal section. The sample size was calculated using Open Epi version 3 software, based on records obtained from the Bureau Agricole Communal (BAC). In addition, the target population is estimated at 1,400 market gardeners, and the survey is conducted among a sample of 113 farmers in the communal section of San-Yago (Figure 15).

3.3.4- Questionnaire

A survey form was drawn up to facilitate data collection, in order to acquire information on farmers' means of production, the techniques they use for their activities, and the evaluation of technical and economic crop profitability indicators (Appendix 1).

3.3.5- Data collection

Data is collected in the field by asking randomly selected market gardeners questions. Once these market gardeners had been located with the help of the BAC and perimeter managers. The questions focused on their means of production, production factors, technical itineraries, production and economic profitability indicators.

3.3.6- Data processing and analysis

Once the data has been collected in the field, it is cleaned, processed and analysed. The following steps were taken for this task:

➤ Analysis of survey forms to identify any errors or missing data;

➤ Question forms are coded to make data entry and subsequent checking easier, by assigning a code number to each completed form.

The database is created in Excel version 2019, then exported to SPSS version

17.0.0. We then exported the data concerning the means of production and the economic profitability indicators to InfoStat version 2018. Monovariate analyses and statistical tests were carried out.

3.3.7- Calculation method

Various methods have been used to estimate vegetable crop production and economic profitability indicators.

3.3.7.1- Vegetable production

To estimate the production of these crops, data were collected on production indices. The product of the number of bags harvested and the weight of one bag. It should be noted that a manmil bag filled with leeks weighs 145 kg, and for the marketing of carrots and beetroot, large batches weigh 50 kg each for these two crops, according to BAC data. Production (Kg) = Number of bags harvested × weight of one bag

3.3.7.2- Estimation of economic calculations

➢ $\mathbf{PB} = \sum_i^n Q_i P_i$

➢ **NPV** =PB-CI-Amortization

➢ **Depreciation**= (n*p)/ D: n: number of units, p: unit price, D: useful life

➢ $\mathbf{VJT} = \dfrac{VAN}{Temps\ de\ travail}$

➢ RA=VAN-MOF-Capital (cost in use or rental value + interest amount)

➢ **Re=RA/CT*100.** Re: economic profitability.

3.3.7.3- Estimate of salaried workforce

The estimate of salaried labour is therefore calculated using the method proposed by Aïhounton et al, (2016), the Total Number (TN) of workers in Human Work Units (HLU) is given by the following formula:

ET = (number of men) + 0.75 * (number of women) + 0.50 * (number of children aged 6 to 14). To convert to man-days (hd), ET was multiplied by the total duration (Td) of the operation (in hours) divided by 5.

3.3.7.4-Categorisation   of vegetable farms in the municipality of Saint Raphaël The main criteria for classifying vegetable farms in the municipality of Saint Raphaël are as follows:

➢ The total area farmed in hectares, a criterion introduced by the MARNDR in 2015 ;
➢ The monetary value of the gross product in HTG, a criterion proposed by INSEE in 2017.
These two criteria were combined for the study, which focuses on the economic profitability of vegetable farms. Four categories of vegetable farms were defined: poor, poor, medium and well-to-do.

3.3.7.5- Statistical tests

Two statistical tests are used to analyse the economic profitability indicators. These tests are described as follows:
Normality test (Shapiro-wilk)

➢ H0: The variables (age, experience, household size, economic rate of return, available labour, area) follow a normal distribution.

➢ H1: The variables (age, experience, household size, economic rate of return, available labour, area) do not follow a normal distribution.

ANOVA test

➢ H0: Socio-economic variables have no influence on the economic profitability of vegetable farms in the commune of Saint-Raphaël.

➢ H1: Socio-economic variables influence the economic profitability of vegetable farms in the commune of Saint-Raphaël.

Pearson correlation test

➢ H0: There is no correlation between the variables (age, number of years of experience, surface area, household size) and the rate of economic profitability of vegetable farms in the commune of Saint-Raphaël.

➢ H1: There is a correlation between the variables (age, number of years of experience, surface area, household size) and the rate of economic profitability of vegetable farms in the commune of Saint-Raphaël.

Rejection condition H_0 $P_{value} \leq 0,05$

IV- RESULTS

This chapter presents the results of the study in detail. Firstly, it describes the characteristics of the farmers and farms. It then classifies the vegetable farms in the area into different categories, such as poor, medium and well-to-do. It then examines production factors and profitability indicators.

4.1- Characteristics of farmers and farms

This section presents the socio-economic characteristics of the farmers, as well as the characteristics of the farms.

4.1.1- Education

According to the farmers' characteristics, the majority of them (79%) have attended school, compared with 21% who have never attended school and are therefore illiterate in their mother tongue (Figure 6).

4.1.2- Sex of respondents

Market gardening is mainly practised by men, who account for 80% of the farmers surveyed, while only 20% are women. This explains why women are more involved in trade (Figure 4).

4.1.3- Main activity of market gardeners

Farming is the main economic activity for 55% of the farmers in the sample, while secondary activities account for 45% of the sample (Figure 5). Trade is the second main activity for farmers in the area, accounting for 19% of the sample.

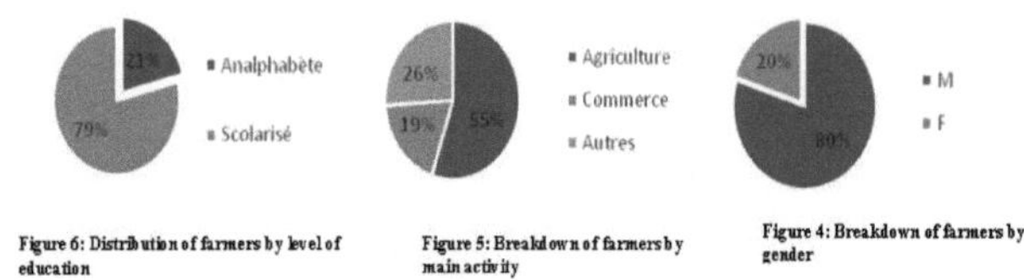

Figure 6: Distribution of farmers by level of education

Figure 5: Breakdown of farmers by main activity

Figure 4: Breakdown of farmers by gender

4.1.4- Marital status of respondents

In the commune of Saint-Raphaël, the majority of farmers are married men and women, representing 50% of our sample, compared with 35% in a common-law union (graph 1).

4.1.5- Destination of vegetables

Over 71% of the farmers in the sample consider farming in the area to be a source of income. However, for less than 27% of farmers, it is both their main source of income and their main source of consumption (Figure 7).

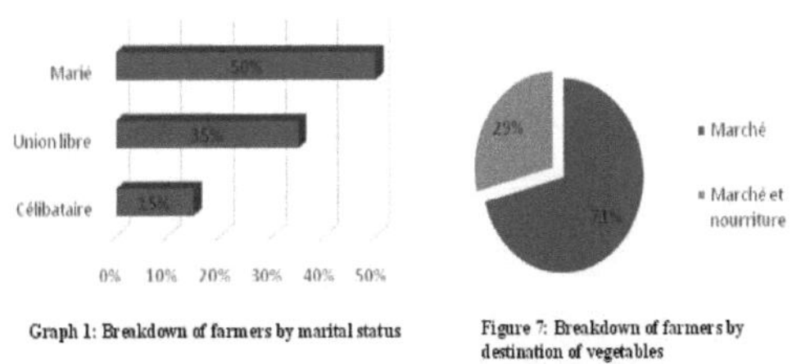

Graph 1: Breakdown of farmers by marital status

Figure 7: Breakdown of farmers by destination of vegetables

4.1.6- Age and experience of respondents

The average age of market gardeners is around 45 ± 10 years, with an estimated average experience of 21 ± 9 years. The most experienced market gardeners in the zone have up to 70 years' experience (table 11).

4.1.7- Source of seed supply

Analysis of the results shows that 71% of the farmers in the sample buy their seed from an input shop, while 29% buy it on the market (table 11).

4.1.8- Agricultural training

According to the results, 38% of respondents had taken part in agricultural training at least once, while more than half of the sample (62%) had not been invited by the authorities to take part in agricultural training (table 11).

4.1.9- Access to credit

Only 20% of the farmers surveyed have access to agricultural credit, while 80% do not. This low percentage of access to credit could be explained by a lack of credit at the BNDA and by the fact that the majority of farmers have not fulfilled the conditions required to obtain credit from the BNDA. On the other hand, a private institution in the area grants agricultural credit at usurious rates and with no grace period, which discourages vegetable farmers from applying (table 11).

4.1.10- Assistance

Statistical analyses of this study show that 40% of the farmers in the sample received technical assistance, while 20% received financial assistance, and only

2% of the sample received material assistance directly from the BAC (table 11).

4.1.11- Membership of a farmers' organisation

According to the results of the statistical analysis, 40% of vegetable growers belong to a farmers' organisation, while 60% have not grouped together or have not familiarised themselves with sharing their experiences (table 11).

4.1.12- Method of sale

Farmers in the commune of Saint-Raphaël have various means of accessing land for market gardening, notably through inheritance, purchase or lease. There are three types of land tenure: direct tenure, where 61% of the land on the farms in the sample is farmed directly, compared with 26% indirectly. On the other hand, in the mixed mode, 13% of the land on the farms in the sample is farmed both directly and indirectly (graph 2).

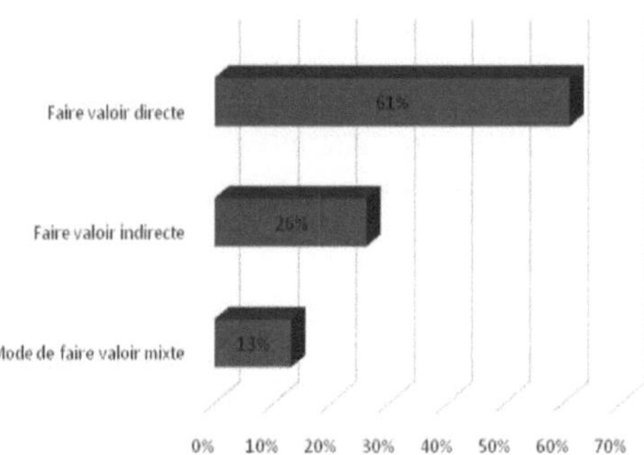

Figure 2: Distribution of farmers by land acquisition method

4.2- Technical itineraries applied to the production of the crops concerned

In the commune of Saint-Raphaël, vegetable production is the second main use of the irrigated perimeters in the communal section of San-Yago. The various stages involved in vegetable production are summarised below.

4.2.1- Nursery

Vegetable plots begin with the preparation of seedlings in the nursery, with the exception of carrots, which are sown directly in beds or flat. For leeks and beetroot, the seedlings remain in the nursery for a month before being transplanted. Preparing the soil in the nursery requires meticulous attention and a lot of manpower for weeding, spraying and watering. Depending on the density of the stand, one or two phytosanitary treatments are applied to the seedlings.

4.2.2- Soil preparation

In the commune of Saint-Raphaël, the soil is prepared for planting vegetables with great precision. It begins with one or two ploughings, either using a plough or animal traction. Next, market gardeners use animal traction to create the beds, a process that is essential to avoid any excess water that could compromise the optimum development of the plants. It should be stressed that this practice varies from one farmer to another.

4.2.3- Transplanting

Transplanting leeks (Allium porum) and beetroot takes place 30 to 45 days after the nursery has been set up. They are planted on either side of previously established beds.

4.2.4- Sowing

Once the soil has been prepared and well harrowed, the farmers sow the seeds of carrots (Daucus carotta) and beetroot (Bêta vulgaris). The seeds are broadcast and the distance between them is not uniform.

4.2.5- Maintenance

Vegetable plots are generally weeded twice, depending on the level of weed development. The first weeding is carried out about three weeks after transplanting and sowing. A second weeding is carried out around 22 days after the first.

4.2.6- Phytosanitary measures

Farmers use locally available products to combat the insects and diseases that attack their crops. The choice of sanitary products varies according to the type of crop and the type of pest. For example, for leeks, they use mancozeb, carbaryl and malathion, while for carrots and beetroot, they opt for malathion, mancozeb, thiophanate-methyl, celcron, ridomil, tricel and others. The number of applications varies according to the level of crop infestation. However, some farmers are finding it difficult to acquire the necessary quantities of products due to financial constraints, which would result in considerable yield losses.

4.2.7- Irrigation

Vegetable crops are often sensitive to excess water. Irrigation frequency is determined by the soil's capacity to retain water. Farmers use supplementary irrigation in the fields to encourage plant development. The frequency of irrigation varies according to the stage of development of the crops; on average, it is carried out twice a week.

4.2.8- Harvesting

Vegetables are harvested between 75 and 90 days after transplanting, depending on the species grown. Leeks are harvested between 75 and 90 days after transplanting, while carrots (Daucus carotta) and beetroot (Bêta vulgaris) are harvested 100-120 days after sowing.

4.2.9- Farm equipment and materials used on surveyed farms

A wide range of equipment is used on vegetable farms, mainly traditional tools. These include watering cans, hoes, machetes and rakes, as well as draught cattle and sometimes tractors. However, there is a shortage of modern equipment in the study area. Only 20% of the farmers in the sample use tractors as a modern means of ploughing and harrowing their plots, while 70% use animal traction, compared with 10% who use traditional tools such as hoes and pickaxes (graph 3).

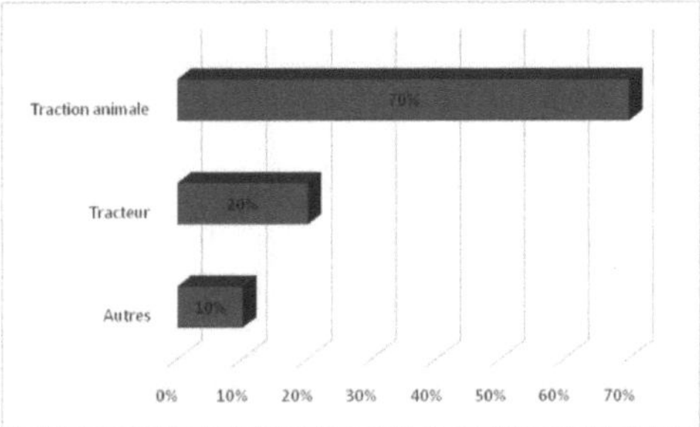

Figure 3: Breakdown of farm equipment used by farmer

4.3- Analysis of production factors for vegetable crops

This section examines the production factors used to grow the crops covered in the study. The statistical analyses are presented in the table below.

4.3.1- Seeds

In the commune of Saint-Raphaël, the results of this study show that farmers use an average of 9.83 ± 1.84 kg/ha, 6.74 ± 9.69 kg/ha and 5.47 ± 8.62 kg/ha respectively for leek, carrot and beetroot cultivation (table 4).

4.3.2- Fertilisers

Fertiliser management remains one of the most difficult agricultural operations to master. In the area studied, the mineral fertilisers used are a combination of two types of formulation: urea (46-0-0) and complete (12-12-20). The quantity of fertiliser used for the same cultivated area varies considerably from one farm to another. The results show that, on average, market gardeners applied 418 ± 216, 235 ± 146 and 237 ± 177 kg/ha, respectively.
fertiliser (table 4).

4.3.3- Intermediate consumption

The cost of intermediate consumption varies considerably from one farm to another for the same cultivated area. On average, market gardeners invest 72982 ± 38,636, 49,480 ± 29,734 and 29,112 ± 18,850 gourdes per hectare.

4.3.4- Organisation of agricultural work in the municipality of Saint-Raphaël

To carry out agricultural activities such as creating beds, planting, sowing, transplanting, weeding and other similar tasks, two types of labour are mobilised: family labour and hired labour, often organised in work groups. Traditional forms of organisation such as mutual aid and coumbite are no longer common in the area. The two types of work used in the area are detailed below.

4.3.4.1- Family workforce

In terms of family labour, each farm has an average of 5 human labour units (HLUs). It is therefore clear that family labour helps to reduce vegetable production costs in the area.

4.3.4.2- Squad

In the commune of Saint-Raphaël, particularly in the communal section of San-Yago, farming activities are carried out in the form of squads, generally made up of groups of men who work on farms as a service, according to the needs of the farmer. In this type of farming, the team leader mobilises his group to carry out the farming tasks, with an average of 19 Man-Days per hectare used for each farming activity.

4.3.5- Area sown to vegetable crops in the study area

For the land factor, the area sown averages 0.61±0.261 ha, and the maximum area does not exceed 2.58 ha (Table 4).

4.3.6- Amount of credit allocated to vegetable growers in the study area

The average amount of credit granted in the zone is 93,214 ±73,306 gourdes, and the minimum amount received is 15,000 gourdes, compared with a maximum of 500,000 gourdes.

4.3.7- Total cost of vegetable crops

In terms of the total cost of vegetable production, market gardeners invest an average of 208,009±116,265, 134,738±79,950 and 100,670±61,264gourdes per hectare respectively (table 4).

Table 4: Presentation of the production factors for the crops covered in the study

Production factor	Obs.	Average	Standard deviation	Variance
Leek seed (kg/ha)	113	9,83E+00	1,84E+01	336,985
Carrot seed (kg/ha)	113	6,75E+00	9,69E+00	93,923
Beet seed (kg/ha)	113	5,47E+00	8,62E+00	74,349
CI Leek (gourdes/ha)	113	72982	38 633	9,45E+09
CI carrot (gourdes/ha)	113	49 480	29 734	6,47E+09
CI beetroot (gourdes/ha)	113	29 112	18 850	2,03E+09
Leek fertiliser (kg/ha)	113	418,21	516,74	106,81
Carrot fertiliser (kg/ha)	113	235,7168	412,86	68,179
Beet fertiliser (kg/ha)	113	237,7418	523,82	109,756
Labour cost C. leek (gourdes/ha)	113	105675	246858	6,09E+10
Carrot labour costs (gourdes/ha)	113	61045,30	118460,46	1,40E+10
Sugar beet labour costs (gourdes/ha)	113	41910,93	88643,34	7,86E+09
Total cost C. leek (gourdes/ha)	113	208 009	116 265	1,28E+11
Total cost C. carrot (gourdes/ha)	113	134 738	79 950	4,94E+10
Total beet cost (gourdes/ha)	113	100 670	61 264	3,05E+10
Agricultural area	113	0.617	0,261	2

4.4- Vegetable crop production

For the three types of vegetable, average production was : 5613.27± 1558, 7498.05± 1796 and 6888.5± 1912 kg for leek, beetroot and carrot respectively (Table 5).

Table 5: Descriptive statistics for production of the three crops

culture	Variable	N	Means.	Var.	Min.	Max
Beetroot	production (Kg)	113	6 888,5	20433584,46	2000	50000
Carrot	production (Kg)	113	7 498,05	23066334,26	3000	55000
Leek	production (Kg)	113	5 613,27	3392124,68	2000	10500

4.5- Classification of vegetable farms in the municipality

Statistical analysis reveals that 11 vegetable farms, or 10% of the farms in the sample, are considered poor. They represent the smallest category, with an average gross annual product of 157,500 gourdes ($1,193) at the BRH reference rate (132 gourdes = 1 dollar) for an average cultivated area of 0.168 hectares under vegetable production. These farms operate without training, assistance or access to agricultural credit. They lack the resources needed to increase their cultivated area and obtain a higher gross product. It should be noted that most of these farmers sell their labour to support themselves.On the other hand, 18 vegetable farms, or 16% of the farms in the sample, are classified as poor, with an average gross annual product of 317,500 gourdes ($2,405) for an average cultivated area of less than 1 hectare, or 0.76 hectares.52 vegetable farms in the communal section of San-Yago, i.e. 46%, are classified as such, making it the most widespread category in the study area, with an average cultivated area of no more than 1 hectare. The average gross annual product is 667,101 gourdes ($5,053). Some farmers in this category have access to agricultural credit and regularly attend agricultural training courses.Lastly, 32 vegetable farms, i.e. 28% of the farms in the sample, are considered to be well off, with an average annual gross production product of 797,500 gourdes ($6,041). vegetables. Their cultivated area does not exceed 2.58 hectares, with an average of 2 hectares. These farms are characterised by significant means of production, they have access to agricultural credits from the BNDA and participate actively in agricultural training.

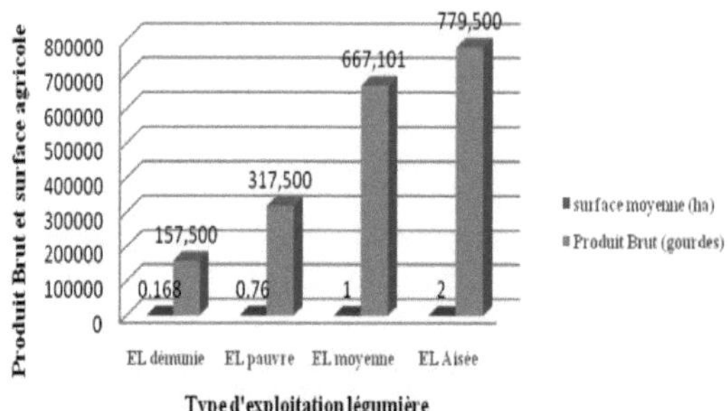

Figure 4: Classification of vegetable farms in the municipality of Saint-Raphaël

4.6- Analysis of economic profitability indicators for the crops covered

This section provides a comprehensive overview of the key economic indicators used to assess the economic profitability of the farms studied.

4.6.1- Gross revenue of vegetable farms

Analysis of the data highlights the average gross income from leek, carrot and beetroot crops, producing 601,257 ± 421,339, 320,580 ± 248,772 and 165,214 ± 147,351 gourdes per hectare respectively (table 6). Vegetable growing in the commune of Saint-Raphaël therefore generates a total gross product of 1,087,053.44 gourdes per hectare.

Table 6: Gross product in GVA per ha of selected vegetable crops

Culture	variable	N	Means.	Min.	Max
Beetroot	PB	113	165 214,54	19 380,00	2 325 581,00
Carrot	PB	113	320 580,96	28 979,00	6 201 550,00
Leek	PB	113	601 257,94	103 359,00	9 689 922,00
Total	1 087 053,44				

4.6.2- Added value of vegetable farms

In terms of value added, vegetable crops generate average wealth of 515,030 ± 385,896, 26,3725 ± 224,156 and 126,986 ± 128,817 gourdes per hectare respectively. (table 7). In total, these crops contribute an estimated 905,743.24 gourdes to wealth in the commune of Saint-Raphaël. It is therefore remarkable to see that vegetable crops play a significant role in wealth creation in this commune.

Table 7: Value added in GVA per ha of selected vegetable crops

Culture	Variable	N	Means.	Min.	Max
Beetroot	VA	113	126 986,52	1 452,00	1 820 439,00
Carrot	VA	113	263 725,87	16 905,00	5 239 044,00
Leek	VA	113	515 030,85	60 509,00	8 634 393,00
Total	905 743,24				

4.6.3- Adding value to the working day on vegetable farms

Average daily yields were 25,405 ± 50,372, 14,355 ± 31,073, and 7,405 ± 21,963 gourdes/day respectively for leek, carrot and beetroot crops in the communal section of San-Yago (table 8). The average total value of vegetable crops was 15,722.11 gourdes per hectare.

Table 8: Value of the working day for selected vegetable crops

Culture	Variable	N	Means.	Min.	Max
Beetroot	VJT	113	7 405,07	45,38	179 105,30
Carrot	VJT	113	14 355,48	565,25	233 368,86
Leek	VJT	113	25 405,79	1 890,89	395 675,06
Avg	15 722,11				

4.6.4- Income from vegetable farms

The average income per hectare from leek, carrot and beetroot crops in the communal section of San-Yago is 393,707 ± 331,922, 197,398 ± 188,099 and 83,662 ± 86,347 gourdes per hectare respectively (table 9). Vegetable growing in the commune of Saint-Raphaël therefore generates an average total income of 674,767 gourdes per hectare.

Table 9: Income in GVA per ha from selected vegetable crops

Culture	Variable	N	Means.	Min.	Max
Beetroot	Income	113	83 662,09	233,00	1 264 884,00
Carrot	Income	113	197 398,15	1 804,00	4 683 488,00
Leek	Income	113	393 707,01	10 848,00	8 001 318,00
Total	674 767,25				

4.6.5- Economic rate of return for vegetable crops

In the communal section of San-Yago, vegetable crops generate distinct rates of return, with percentages varying according to the type of crop. For leeks, the rate is 61%, while for carrots and beet the rates are 48% and 40% respectively (table 10). Overall, the average profitability for these crops is 50%.

Table 10: Economic rate of return for selected vegetable crops

Culture	Variable	N	Means.	Min.	Max
Beetroot	Profitability	113	40%	2%	90%
Carrot	Profitability	113	48%	13%	90%
Leek	Profitability	113	61%	7%	99%

4.7- Analysis of socio-economic factors influencing profitability

The results of the ANOVA and Pearson tests revealed the significant influence of various socio-economic factors on the economic profitability of vegetable crops in the commune of Saint-Raphaël.The ANOVA analysis showed that access to agricultural credit, participation in agricultural training, farmers' level of education and their involvement in at least one agricultural extension activity had a significant impact on economic profitability, with a significance level of 1% (p = 0.0001) (see Figures 8 to 10). On the other hand, the gender of the farmers and their membership of a farmers' organisation did not show any significant influence on the economic profitability of vegetable farms. In addition, the Pearson test revealed significant correlations between the number of years of experience of vegetable growers, their age, household size and the rate of economic profitability of vegetable farms. On the other hand, no significant correlation was observed between available labour and area sown and economic profitability (Figure 11).

4.8- Analysis of constraints linked to the crops covered in the study

The constraints encountered in the commune of Saint-Raphaël for the production of the crops under consideration are multiple, encompassing technical, environmental, economic and financial aspects that compromise their economic profitability.

4.8.1- Technical constraints

In the commune of Saint-Raphaël, the technical constraints affecting vegetable crops are particularly marked. The use of out-of-date or poor-quality seeds seriously compromises crop yield potential. In addition, due to a lack of supervision and training, market gardeners often adopt unsuitable cultivation practices and crop combinations that are not recommended, which can be detrimental to the growth and health of the plants. Irrational and inappropriate use of pesticides and fertilisers also exacerbates these problems, leading to ineffective nutrient and disease management. In addition, the lack of sufficient modern agricultural equipment severely limits producers' ability to improve their farming practices and optimise their production. These technical constraints are holding back the development of vegetable crops in the region, reducing their profitability and competitiveness.

4.8.2- Environmental constraints

The environment has a considerable influence on farming activities, and vice versa. In the commune of Saint-Raphaël, periods of drought due to the frequent lack of rain, combined with a poorly laid-out irrigation system blocked by sediment, reduce the flow of water available for irrigating crops, while the use of drainage water for irrigation accelerates the spread of disease and insects from one plot to another, and also poses a threat to market gardening in the area. In addition, extreme and unpredictable weather events significantly affect market garden production in the commune. As a result, these constraints delay planting periods, thus affecting the crop calendar and the availability of vegetables on the market and making long-term planning difficult.

4.8.3- Economic and financial constraints

Economic and financial constraints are the most worrying challenges facing market gardeners in the commune of Saint-Raphaël. In less than a year, the price of inputs has risen considerably: for example, a 100lb bag of complete fertiliser that used to cost between 4,000 and 5,000 gourdes is now sold for 10,000 gourdes. Similarly, the price of some crop protection products has risen significantly, from 350 gourdes to 750 gourdes or more. The ever-increasing cost of seeds and labour, combined with low self-financing capacity and limited access to agricultural credit, is limiting farmers' ability to make full use of their plots, further hampering vegetable production in the area.

4.8.4- Infrastructure constraints

Infrastructural constraints, such as the lack of equipment for post-harvest storage and processing of vegetables, mean that market gardeners have technical difficulties in ensuring the durability of their produce, which considerably increases post-harvest losses and reduces their income. In addition, the lack of investment in farm mechanisation and the agri-food industry limits production and processing capacity, thereby hampering the growth potential of the commune's vegetable sector.

4.8.5- Institutional constraints

Institutional constraints have a considerable impact on the economic profitability of vegetable growing in the commune of Saint-Raphaël. The lack or absence of financial and technical support, particularly in terms of subsidies and training, increases production costs and limits the adoption of modern practices and technologies, which could reduce vegetable crop yields in the municipality. What's more, the lack of adequate infrastructure for storing, preserving, perpetuating, transporting and marketing vegetables leads to post-harvest losses and low selling prices, reducing market gardeners' incomes. The absence of clear agricultural policies and of appropriate regulations creates a climate of uncertainty, discouraging investment and exposing farmers to losses due to uncontrolled pests and diseases. Finally, limited access to public services, such as research and training programmes, hampers innovation and the improvement of farming practices in this commune. Together, these institutional constraints compromise the economic profitability of vegetable farms by reducing their ability to maximise technical profitability (production, yields), manage allocative profitability (efficient costs) and access markets competitively.

4.9- Analysis of assets

The municipality of Saint-Raphaël has three irrigated areas, two of which are located in the San-Yago section of the municipality: the large Merlaine area and the small Merlaine area. In addition, there are more than 40 untapped springs in Merlaine. Exploiting these opportunities simply requires a popular and public will to redevelop the existing systems and increase the capture of springs at Merlaine in order to increase the quantity and flow rate of the small Merlaine perimeter. In addition, the area's soil and climate conditions are ideal for vegetable production. The soil provides an optimal environment for market gardening. In addition, the presence of farmers' organisations, the BAC and young agronomists and agricultural technicians in the commune is a considerable asset. These players represent a dynamic human capital that market gardeners and decision-makers can mobilise to improve the technical and economic profitability of the crops studied, and even explore new opportunities.

By harnessing these resources effectively and encouraging collaboration between the various stakeholders, it is possible to turn these assets into tangible realities, thereby helping to strengthen the local economy and promote the sustainable development of the municipality via the vegetable sector.

4.10- DISCUSSIONS

The aim of this research is to analyse the economic profitability of vegetable farms in the commune of Saint-Raphaël. With regard to the characteristics of the farmers, it should be noted that vegetable growing in this commune is largely dominated by men, representing 80% of the sample. This finding is not far removed from the results obtained by St-Pierre, Jean Luc (2022) in the communal section of Grand Fond, commune of Kenscoff, where 85% of the farms in his sample were run by men. This predominance of men could be explained by the role traditionally attributed to women in the marketing of agricultural products, while farming remains mainly a male activity. The results of the study also reveal that 21% of the farmers in the sample are illiterate, a slightly higher percentage than the 13% obtained by St-Pierre, Jean Luc (2022) in the commune of Kenscoff. This difference can be attributed to the geographical location of Saint-Raphaël, which is further from urban centres and therefore less accessible to educational institutions. In addition, 38% of the farmers in the sample are members of a farmers' organisation, while only 18.5% are in the commune of Kenscoff. This difference can be explained by the better organisation of farmers in the commune of Saint-Raphaël. The average age of farmers is 45, which is not very far from the 43 reported by St-Pierre Jean Luc (2022). This slight difference could be due to the fact that Kenscoff is a peri-urban area where the market for vegetable produce is more accessible, thus attracting a younger population to vegetable farming. Furthermore, the average size of vegetable farms in the commune of Saint-Raphaël is relatively small, at 0.617 hectares, compared with 0.76 hectares in the commune of Kenscoff. This disparity can be explained by the level of demand and incentives in each region; Kenscoff being closer to the metropolitan area, demand for vegetables is potentially higher. In the communal section of San-Yago, two types of labour are mainly used by market gardeners: hired labour organised into squads, and family labour. Family labour is mainly used for simple tasks, while hired labour is mobilised for more demanding tasks such as weeding and spreading fertiliser. To meet the labour requirements of market gardening, farmers mobilise an average of 19 man-days per hectare for various activities such as transplanting, sowing, hoeing and spreading fertiliser. As far as seeds are concerned, in the commune of Saint-Raphaël, sowings vary depending on the crop. For leeks, a high quantity of seed is used per hectare, far exceeding the recommendations in the literature, which range from 1.5 to 4 kg/ha. This difference could probably be explained by failure to observe the recommended planting distances. Market gardeners sow a slightly higher quantity of carrots, around 6.74 kg/ha, compared with the recommendations of 4 to 5 kg/ha. However, for beetroot, market gardeners used less seed than recommended, sowing around 5.47 kg/ha, whereas recommendations are generally between 8 and 10 kg/ha according to the literature.From an economic point of view, market gardeners incur very high production costs. To produce 1 hectare of leeks, the farmer invests 208,009 gourdes, while for carrots and beetroot, the costs are 134,738 and 100,670 gourdes respectively. Despite the high contribution of family labour, the average total cost of production for these crops remains very high, at 443,417 HTG ($3,359) per hectare.Vegetable growing generates an average total gross product of 1,087,053 HTG (8,235 USD) per hectare, surpassing the results reported by St-Pierre-Jean Luc (2022) for the commune of Kenscoff, where vegetable growing produced an average gross product of 7,710 USD for the main crops considered. This disparity could be explained by the fact that the main crops grown in the commune of Saint-Raphaël differ from those grown in Kenscoff. Despite the fact that farmers in Saint-Raphaël are generally older and farm smaller areas than those in Kenscoff, vegetable farms in Saint-Raphaël recorded a higher gross product in the same study year.The analysis also reveals that market gardening generates wealth valued at USD 6,861 in the commune of Saint-Raphaël, compared with USD 6,277 in Kenscoff. This difference can be explained by intermediate consumption of the main vegetable crops in the commune of Kenscoff, which is three times higher than that of vegetable crops in the commune of Saint-Raphaël. commune of Saint-Raphaël. In addition, the main crops covered are not identical in the two communes. The results are also revealing when it comes to the value of the working day. On average, a day's work growing vegetables in the communal section of San-Yago is valued at 15,722.11 HTG (119 USD) per hectare per day. In other words, for one hectare of land, a day's work growing vegetables is valued at 15,722.11 HTG for the year 2022, after deduction of intermediate consumption and depreciation. These figures underline the vital importance of vegetable growing in creating wealth in the municipality of Saint-Raphaël. In addition, the average total income generated per hectare of vegetable crops is 674,767.25 HTG ($5,111) in the commune of Saint-Raphaël, compared with 9,842,000.00 FCFA ($16,200.00) in the communes of Imanan and Tagazar in Niger. This difference highlights the disparities in wealth between regions, as well as the economic, agri-environmental and structural factors that influence the profitability of vegetable farms.The economic rate of return on vegetable production in the commune of Saint-Raphaël is 50% on average. In other words, for every 100 gourdes invested in market gardening in the communal section of San-Yago, market gardeners obtain an income of 50 gourdes. However, it should be noted that leek growing seems to be more profitable in this commune.The results of the ANOVA test revealed a significant dependence of economic profitability on several socio-economic variables of the farmers. In particular, access to agricultural credit, farmers' level of education and participation in agricultural training all had a significant influence on economic profitability, with a significance level of 1% (p = 0.0001). These results suggest that access to agricultural credit can increase the rate of profitability, and that a higher level of education among farmers is associated with increased farm profitability. Furthermore, the results of the Pearson test showed that household size exerts a significant and positive influence on the economic rate of return on vegetable crops, with a p-value of 0.0103. This highlights that, despite the predominance of salaried labour in the region, household size remains a crucial factor, providing an important source of unsalaried labour for vegetable production in the commune of Saint-Raphaël. In addition, the experience of market gardeners also showed a significant and positive influence, with a p-value of 0.0001. This positive influence can be attributed to the growers' ability to correct technical errors from previous years, thereby improving their economic profitability. On the other hand, the age of the market gardeners showed a significant but negative influence on the rate of economic profitability, suggesting that an increase in age could be associated with a decrease in profitability. These findings confirm our first research hypothesis. With regard to the constraints hindering the economic profitability of vegetable growing in the commune, the results reveal technical, environmental, economic, financial, infrastructural and institutional constraints. In the light of these findings, we concluded that the constraints facing vegetable growers in San-Yago are no different from those encountered in other parts of the country, where farms are left to fend for themselves, with no new production techniques, no new equipment, no support and almost no agricultural credit. These findings confirm our final research hypothesis.

CONCLUSION AND RECOMMENDATIONS

In this study, we propose to analyse the economic profitability of vegetable farms in the municipality of Saint-Raphaël in the year 2022, with a focus on the factors influencing the economic profitability of vegetable crops and the constraints to vegetable production in this area. To this end, specific objectives were defined in order to answer these questions.Firstly, statistical analysis showed that 80% of respondents were men, and 55% of them practised farming as their main activity. Only 20% had access to agricultural credit, and the average area cultivated with vegetables was 0.617 hectares. For our second specific objective, vegetable farms were categorised into four groups: poor (10%), poor (16%), medium (46%) and well-to-do (28%).For our third objective, we assessed the economic profitability indicators. Vegetable growing generated an average total gross product of 1,087,053 gourdes and an average total income of 674,767.25 gourdes per hectare. Economic rates of return vary from crop to crop, with the average for vegetable farms estimated at 50%. In the light of these results, it was concluded that vegetable farms in the commune of Saint-Raphaël are economically profitable. However, the results of this study show that leek growing stands out statistically as being more profitable from an economic point of view. At the same time, the ANOVA and Pearson tests highlighted the factors influencing the economic profitability of the crops studied, which are mainly of a socio-economic nature, such as household size, farmers' level of education, access to agricultural credit and others.According to the last objective, various constraints have been identified, particularly of a technical, environmental, financial and infrastructural nature. Despite these constraints, the municipality has considerable assets. Favourable soil and climate conditions and the human capital available in the region provide a favourable environment for the development of market gardening. To exploit these opportunities, however, we need to strengthen our resolve and take concrete action to turn them into a tangible reality. After analysing the various problems and constraints facing the Haitian agricultural sector, particularly in the commune of Saint-Raphaël, it appears that farms could be more profitable. Indeed, environmental factors and certain favourable social factors are present in the area. However, the lack of enlightened policies for the agricultural sector at national level has led to a lack of incentives and attractive conditions to encourage investment in this sector. It is crucial to stress that there can be no agricultural development without farm development, and no farm development without improved conditions for farmers. Similarly, food self-sufficiency cannot be achieved without financial autonomy at farm level, because agricultural development is intrinsically linked to farm development. At the end of this study, the research hypothesis that socio-economic factors influence the economic profitability of vegetable crops in the commune was confirmed. However, this study has certain limitations and did not take into account all the factors that could explain this phenomenon. It would therefore be appropriate for other researchers to examine this question in the same region and for the same crops in order to identify other factors that have a positive and significant influence on the economic profitability of vegetable farms. So, to improve the economic profitability of vegetable farms in the area, short- and medium-term recommendations are made:

In the short term, decision-makers must :
➢ Focusing on access to agricultural credit in the commune of Saint-Raphaël;
➢ To train vegetable growers in the rational and optimal use of mineral fertilisers and pesticides, and to facilitate access to quality inputs;
➢ Encourage farmers to set up associations, in particular by providing financial and material assistance;
➢ Create attractive conditions to make it easier for young people to enter the industry; encourage farmers to grow more leeks.
In the medium term, decision-makers must :
➢ Set up an agricultural policy aimed at: intensifying crop areas, controlling prices and subsidies while developing agricultural mechanisation and monoculture;
➢ Strengthen the capacity of producers, particularly in technical terms;

➢ Redeveloping and increasing the hydro-agricultural infrastructure in the area, this policy would enable an average improvement in the area's level of profitability;
➢ Increase the financial knowledge and entrepreneurial skills of market gardeners;
➢ Improving the functioning of markets and trading systems ;

➢ To create the necessary conditions to facilitate the establishment of agri-food industries in order to stimulate, enhance and sustain vegetable production in the area.

BIBLIOGRAPHICAL REFERENCES

1. **Adamou, I. (2020).** Production techniques for irrigated crops (Carrot). Niger Ministry of Agriculture [Online]. Accessed 1 April 2024. < https://duddal.org/files/original/88c8be5be533c4e041b455eea6fcc57a6c480bdb.pdf >

2. **Agreste.(2013).**Thefarms légumièresen IIe-de-France.ATE-Avenir Télématique[On online]. Accessed at on 8 August 2023. <http://sgproxy02.maaf.ate.info/IMG/pdf/dossier16_chapitre3.pdf >

3. **Agridea (2017).** Organic farming. [Online]. Accessed April 1, 2024.

<https://www.agridea.ch/fileadmin/AGRIDEA/Theme/Productions_vegetales/AgricultureBiological_Technical_Sheets_Samples/Growing_Crops/4.3.1-10_Betterave.pdf >

4. **Albouchi, L., & Bachta, M. (2007).** Estimation and decomposition of the economic efficiency of irrigated areas to better manage existing inefficiencies. [On-line]. Accessed 7 November 2023. <https://hal.science/cirad-00193606/document >

5. **World Bank. (2021).** Agriculture and Food. World Bank [Online]. Accessed 8 August 2023. <https://www.banquemondiale.org/fr/topic/agriculture/overview>

6. **Barbacar, F., Sadibou, S, Mamadou, D-F., & Bachir, W (2020).** Agroeconomic performance of Super Granulated Urea: the case of rice in Senegal. European Scientific Journal [Online].   364-380.   Accessed on 8 August 2023.<https://eujournal.org/index.php/esj/article/view/12937 >

7. **Bognini, S. (2010).** Market gardening and food security in rural areas. Master's thesis, University of Ouagadougou [Online]. Accessed 30 November 2022. <https://www.memoireonline.com/02/12/5258/m_Cultures-maraicirccheres-et-securite- alimentaire-en-milieu-rural4.html >

8. **CIRAD. (2009).** Economic calculations with Olympe software for reference farm networks and definitions for the PAMPA project. Centre de coopération Internationale en Recherche Agronomique pour le Développement [On line]. Accessed 4 April 2024. <https://agritrop.cirad.fr/563699/ >

9. **Durant, D. (2005).** La rentabilité des entreprises: une approche à partir des comptes nationaux. Bulletin de la Banque de France N° 134. Autorité de Contrôle Prudentiel et de Résolution [On line]. Accessed on 26 November 2023. <https://acpr.banque-france.fr/fileadmin/user_upload/banque_de_france/archipel/publications/bdf_bm/

etudes_b df_bm/bdf_bm_134_etu_2.pdf >

10. **El Ouaamar,S.,Tillie,P., Sanou,F-L., Trèves,V., Girard,C., gomez-Y-Paloma,S., & Cochet, H. (2019**). economic performance of family, employer and enterprise agriculture. Côte d ' Ivoire [Online]. Accessed 26 November 2023.<https://publications.jrc.ec.europa.eu/repository/bitstream/JRC116258/jrc11 6258_online. pdf>

11. **FAO. (1995).** Definitions and concepts. Food and Agriculture Organization of the United Nations. and Agriculture.N 5°[On online]. Accessed at on 23 January 2023.
<https://www.fao.org/3/az958f/az958f.pdf >

12. **FAO. (2000).** The role of agriculture in the development of the least developed countries and their integration into the world economy [On line]. Accessed 23 January 2023.
<https://www.fao.org/publications/card/fr/c/474d46ee-a35c-5454-b693-025e269f18e6/>

13. **FAO. (2001).** Farming systems and poverty: improving farmers' livelihoods in a changing world. Food and Agriculture Organization of the United Nations [On line]. Accessed 23 January 2023.
<https://www.fao.org/3/Y1860f/y1860f00.htm >

14. **FAO. (2004).** Fruit and vegetables for health. Food and Agriculture Organization and Agriculture [On online]. Accessed at on 1 October 2023
<https://www.fao.org/fileadmin/templates/agphome/documents/horticulture/WHO/KOBE-Fran%C3%A7ais.pdf >

15. **FAO. (2022**). The country at a glance. Food and Agriculture Organization of the United Nations [Online]. Accessed 1 April 2023. Retrieved from<
https://www.fao.org/haiti/fao-en-haiti/le-pays-en-un-coup-doeil/fr/ >

16. **FAOSTAT. (2022**). Data on vegetables in Haiti [Online]. Accessed 25 November 2022. <https://www.fao.org/faostat/fr/#data/QCL>

17. **IFAD. (2021).** country strategy note, note Main report and. International Fund for Agricultural Development [Online]. Accessed 10 January 2023 <
https://www.ifad.org/documents/38711624/39485439/Haiti+Country+Strategy+Note+202 2-2023.pdf/43429f0f-05af-751c-1c84-3e0a00958240?version=1.1&t=1652704140286&download=true >

18. **François, P. (2008**). The cropping system: a meaningful concept for thinking about the future. Cahiers Agricultures.Vol.17 N 3 [On online].Accessed at on 1 April 2024, <https://revues.cirad.fr/index.php/cahiers-agricultures/article/view/30717 >

19. **François, R. (2017).** Technico-economic evaluation of leek (Allium porum)

cultivation in the San-Yago section during the 2014-2015 period (case of the large irrigated perimeter, SCIPA), St Raphael [Online]. Bachelor's thesis, Université Chrétienne du Nord.   Accessed on 1 December 2022. <https://issuu.com/reginaldfrancois7/docs/memoire_reginald_bon_document_lisan >

20. **Gras, R. (1990).** Cropping systems, definitions and key concepts. ln Les systèmes de culture. Coord. Combe L., Picard O., (coords.). Paris, INRA, p. 7-14.

21. **Guen, A. (2016).** prendre en compte les enjeux économiques des exploitations agricoles dans les démarches de protection des captages. paris: Ministère de l'Agriculture et de l'Alimentation.

22. **IHSI (2015).** Total population, aged 18 and over; households and estimated densities in 2015. www.haiti-now.org [Online]. Accessed 7 November 2022 <https://www.haiti- now.org/wp-content/uploads/2020/04/population-total-18-and-older-2015.pdf >

23. **ITCMI. (2022).** Fiche techniques valorisées des cultures maraichères et industrielles. Ministère de l'Agriculture et du Développement Rural [On line]. laboress-afrique.org. Accessed at on 1 April 2024 **<https://itcmi-dz.org/wp-content/uploads/2022/06/POIREAU pdf** >

24. **Jean-Denis, S. (2015).** Cours de cultures maraîchères, Damien. Port-au-Prince: FAMV,P. 22.

25. **Jeanniton, J., & Bellande, A. (2016).** HAITI: Plan National d'Investissement Agricole [Online]. Ministry of Agriculture, Natural Resources and Rural Development. Retrieved on 25 November 2022. **<https://www.gafspfund.org/sites/default/files/inline-files/7.%20Haiti_Investment%20Plan**.pdf>

26. **Jouve, P. (2003).** Cropping systems and spatial organisation of territories: a comparison between temperate and tropical agriculture. Proceedings of the international symposium (pp. 7-14). Montpellier [On line]: Accessed on 5 November 2022.<https://agritrop.cirad.fr/518697/1/ID518697.pdf >

27. **Lassana,T.,Konipo,O., & Diagne,A (2021, December**). Analyse de la Rentabilité Économique et Financière de la Production Cotonnière au Mali. Revue Scientifique biannuelle de l'Université Ségou [On line], pp. 108-132. Accessed on 8 August 2023.<https://hal.science/hal-03147509/document >

28. **Laurent, C., & Rémy, J. (2000).** L'exploitation agricole en perspective [On line]. In courrier de l'environnement de l'INRA (N° 41), 5-22. Accessed on 25 November 2022. < **https://hal.science/hal-01220445/file/C43oepe%20-%20copie.pdf** >

29. Malézieux, E., & Trébuil, G. (2000). Agronomy and environmental and natural resource management at CIRAD [On line]. Accessed on 1 December 2022.
<https://www.researchgate.net/publication/252320838_L'agronomy_and_env ironmental_and_natural_resources_management_at_Cirad_Reflections_pro posalselements_of_perspective >
30. Malla, I. A., & Yabi, A. (2023). Les déterminants de la rentabilité économique des entreprises paysannes en milieu rural dans le Borgou au Bénin. African Scientific Journal. Vol 3, N16 [Online]. Accessed 7 November 2023 **<https://hal.science/hal- 03961554/document >**
31. MARNDR (2015). Diagnostic des systèmes de production en vue de la relance de la vulgarisation agricole [Online]. Ministry of Agriculture, Natural Resources and Rural Development.   Accessed on 25 November 2022. **<https://agriculture.gouv.ht/view/01/IMG/pdf/rapport-final- etude_systemes_de_production-2.pdf >**
32. MEF. (2015). programme de développement régional de la boucle centre-Artibonite [Online].Retrieved on 25 November 2022.
<https://www.mtptc.gouv.ht/media/upload/doc/publications/Aposter31072.pdf>
33. **MEF. (2021).** Rural accessibility and resilience project [Online]. Accessed November 25, 2022.
<https://www.mtptc.gouv.ht/media/upload/doc/publications/PARR- BCA-PGES_Centre_Entretien_routier_Saint_Michel_sept2023.pdf >
34. Moïse, P. (2017). Market study of the Port-au-Prince metropolitan area: first step to the establishment of a market garden production farm in Dumé, croix-des- bouquets. Dissertation, Université de Liège [On line]. Accessed 8 August 2023.
<https://matheo.uliege.be/handle/2268.2/3087?locale=fr>
35. **NAZA. (2016, December 31).** Merra-2 satellite-era reanalysis. Retrieved from en.weatherspark.com [Online]. Retrieved August 8, 2023:< https://fr.weatherspark.com/y/25378/Météo-moyenne-à-Saint-Raphaël-Haiti-tout-au-long- of-year>
36. Pirou, J. (2005). Mesure de la rentabilité économique des entreprises. Financial analysis course,University IbnZohr University [On online]. Accessed at on 20 December   2022.
<https://www.studocu.com/row/document/universite-ibn-zohr/analyse-financiere/la- rentabilte-introduction/85050963 >
37. Statista (2023, January 23). Vegetable production by continent 2021. Retrieved from en.statista.com [En line]. Retrieved from on 1 April 2024:
<https://fr.statista.com/statistiques/565143/production-de-legumes-a-l-

echelle- mondiale-par-region/ >

38. St-Pierre, J. (2022). Contribution au maraichage périurbain par l'analyse de la production maraichère dans la commune de Kenscoff (Haïti) : Cas de la section communale de Grand Fond [Online]. Master's thesis, University of Liège. Accessed on 25 November 2022.
<https://matheo.uliege.be/bitstream/2268.2/16318/4/TFE_Jean%20Luc%20S T- PIERRE_s214934.pdf >

39. Sebillotte, M. (1993). Système de culture. Encylopédia Universalis p. 558-961

40. Yehouenou, L. S. (2011). financial profitability of apple cabbage and chilli production under anti-insect netting. Dissertation, UNIVERSITE D'ABOMEY-CALAV [Online]. Accessed 30 November 2022.
<https://agritrop.cirad.fr/569276/ >

APPENDIX

Appendix 1: Survey questionnaire Section 1: General information

N0	Question	Response
I01	Question form number	
I02	Date of interview	
I03	Last name and first name of respondent	
I04	Chief operating officer	
I05	Municipality/section	
I06	Sex of respondent (0=male, 1=female)	
I07	Age of respondent	
I08	Marital status (0=single, 1=married, 2=divorced, 3=common-law)	
I09	Level of education (0=non-literate, 1=primary 2=secondary, 3=post secondary)	
I10	Household size	
I11	Gender and age of the family MO. Rep.	
I12	Main occupation (0=Farming, 1=shopkeeper, 2=teacher, 3=employed) agricultural, 4= bricklayer, 5= other)	

Section 2: technical information

N0		Response
1	Years of experience	
2	Method of acquiring land (0=inheritance, 1=rental, 2=purchase, 3=lease),4=lease and purchase, 5=inheritance, lease and purchase)	
3	Estimated cultivated area in cx	
4	Type of crop	
5	Associated crop (0=No, 1=Yes)	
6	Irrigated area (0=No, 1=Yes)	
	Frequency of watering/irrigation (2 times/month=1; 3 times/month=2; 4 times/month=3)	
7	Seed supply (1=input office, 2=market, 3=office andmarket)	
8	Bioaggressors observed (0= No, 1=Yes)	
9	Equipment used (1= ox plough, 2= tractor, 3= plough and tractor)	
10	Access to credit (0=No, 1=Yes) if yes, the amount received Rep :	
11	Agricultural training (0=No, 1=Yes)	
12	Membership of a farmers' organisation (0=No, 1=Yes)	
13	Contact with a research department (0=No, 1=Yes)	
14	Contact with extension agents (0=No, 1=Yes)	
15	Type of assistance (0=none, 1=technical, 2=financial, 3=material)	
16	Destination of harvest (1=market, 2=household consumption, 3=both)	

Section 3: Technical itinerary

Operations	Culture	Quantity	Period
Nursery preparation and sowing	Leek		
	Carrot		
	Beetroot		
Soil preparation	Leek		
	Carrot		
	Beetroot		
Harrowing and levelling	Leek		
	Carrot		
	Beetroot		
Watering/irrigation	Leek		
	Carrot		
	Beetroot		
Transplanting	Leek		
	Carrot		
	Beetroot		
Fertilisation	Leek		
	Carrot		
	Beetroot		
Spraying	Leek		
	Carrot		
	Beetroot		
Weed control	Leek		
	Carrot		
	Beetroot		
Harvest	Leek		
	Carrot		
	Beetroot		

Section 4: cost

N0	Inputs	Designation	Unit	Culture	Quantity	Ct U	Ct total
1	Earth	An					
	Intermediate consumption						
				Leek			
	Type of seed			Carrot			
				Beetroot			
3				Leek			
	Pesticides			Carrot			
				Beetroot			
				Leek			
4	Fertilizers			Carrot			
				Beetroot			
6	Equipment and equipment			Leek			
				Carrot			
				Beetroot			
Expenses							
		HJ		Culture	Qty weeding	Prices	Cost total
		Hj		Leek			
	Salaried MO	Hj		Carrot			
		Hj		Beetroot			
				Culture	Qty sowing/repi		
		Hj		Leek			
	Salaried MO	Hj		Carrot			
		Hj		Beetroot			
OTHER COSTS							
9				Leek			
	Interest			Carrot			
				Beetroot			
	Ground rent paid to the owner			Leek			
				Carrot			
				Beetroot			
				Leek			
10	Transport			Carrot			

			Beetroot			
			Leek			
11	Fuel		Carrot			
			Beetroot			
	Irrigation fee		Leek			
			Carrot			
			Beetroot			

Section 5: output

Product	Var	Culture	Quantity	Sales PX	Destination of the harvest
		Leek			
		Carrot			
		Beetroot			

What are the constraints encountered in this sector?

Annex 2: Statistical results

Table 11: Characteristics of farmers and farms

Variable		Absolute frequencies	Relative frequencies	Variable	Absolute frequencies	Relative frequencies
	Supply of seed			Training Agricultural		
Shop input		80	71%	No	70	62%
Walk		33	29%	Yes	43	38%
Total		113	100%	Total	113	100%
	Access to credit			Farmers' organisations		
No		90	80%	No	54	52%
Yes		23	20%	Yes	49	48%
Total		113	100%	Total	113	100%
Experience				Assistance		
[2-10]		30	27%	Techni that	45	40%
[11-18]		23	20%	Financi era	23	20%
[19-and more]		60	53%	Materie lle	2	2%
Total		113	100%	No	43	38%
				Total	113	100%
Age						
[17-37]		25	22%			
[38-48]		70	62%			
[58yearsan dmore]		18	16%			
Total		113	100%			

```
Acces-credit

    Variable    N    R²  Adj R²  CV
  Acces-credit 113 0,95   0,92 23,99

  Analysis of variance table (Partial SS)
   S.V.    SS    df   MS    F    p-value
  Model. 26,28   42 0,63 32,85 <0,0001
  ref    26,28   42 0,63 32,85 <0,0001
  Error   1,33   70 0,02
  Total  27,61  112
```

Figure 8: ANOVA test on economic profitability and access to agricultural credit

```
niveau-educ.

    Variable     N    R²  Adj R²  CV
  niveau-educ. 113 0,90   0,83 35,29

  Analysis of variance table (Partial SS)
   S.V.     SS    df   MS    F    p-value
  Model. 127,02   42 3,02 14,35 <0,0001
  ref    127,02   42 3,02 14,35 <0,0001
  Error   14,75   70 0,21
  Total  141,77  112
```

Figure 9: ANOVA test on profitability and level of study

```
For. Agricole

    Variable      N    R²  Adj R²  CV
  For. Agricole 113 0,77   0,63 58,18

  Analysis of variance table (Partial SS)
   S.V.    SS    df   MS    F   p-value
  Model. 21,74   42 0,52 5,61 <0,0001
  ref    21,74   42 0,52 5,61 <0,0001
  Error   6,46   70 0,09
  Total  28,19  112
```

Figure 10: ANOVA test on economic profitability and participation in training courses

```
Correlation coefficients


Pearson correlation

Variable(1)  Variable(2)  n   Pearson p-value
ref          ref          113    1,00 <0,0001
ref          age          113   -0,59 <0,0001
ref          T-menage     113    0,24  0,0103
ref          anne-exper   113    0,82 <0,0001
ref          surf-ha      113    0,05  0,5837
ref          Wd-UTH       113   -0,09  0,3385
```

Figure 11: Pearson test of the influence of quantitative variables on the economic profitability of the zone

Appendix 3: List of photos taken during the survey

Photos 2: vue d'une plot leek de

Photos 1: Irrigation of leeks in association with Beetroot

Photos 6: vue d'une plot de carrot

Photo 3: data collection from a farmer working in the BAC

Photo 5: Sale of carrots at the modern market in Saint-Raphaël.

Photo 4: Sale of beetroot at the modern market in Saint-Raphaël.

Printed by Books on Demand GmbH, Norderstedt / Germany